This book dedicated to-

HALDIA INSTITUTE OF TECHNOLOGY

PREFACE

This book addresses Industrial manufacturing process and Non-traditional & traditional machining process. It is designed for use as a primary handbook for material/mechanical/Industrial engineering course, and as a resource for academic and practical approach.

The main characteristics of this book are:
- It presents modern industrial tour and its operation in practical approaches. The machining process and Modern manufacturing system are discussed in academic way.

CONTENTS

1 PROCESS OF MANUFACTURING
- 1.1. INTRODUCTION
- 1.2. STEEL AND INTEGRATED STEEL PLANT
- 1.3. SAFETY MEASURES & PREVENTION
- 1.4. PLANT OVERVIEW
- 1.5. SINTER PLANT
- 1.6. RMHP
- 1.7. COKE OVEN PLANT
- 1.8. BLAST FURNACE
- 1.9. STEEL MELTING SHOP
- 1.10. MERCHANT MILL
- 1.11. WHEEL AND AXLE PLANT

2 PROCESS OF MACHINNING
- 1.12. INTRODUCTION
- 1.13. DRILLING MACHINE
- 1.14. GRINDING MACHINE
- 1.15. LATHE MACHINE
- 1.16. .CNC EDM
- 1.17. MILLING MACHINE

3 INTELLIGENT MANUFACTURING SYSTEMS

CHAPTER-1

PROCESS OF MANUFACTURING

INTRODUCTION

To manufacture of industrial product,it is essential to implement for better design,optimization,control,operation of chemical, physical, and biological processes, known as process engineering.

Typically, it is a conversion process for final product from raw material.

To find the easiest process for industrial design to final product, manage the complex system as well as utilization of system thinking possibilities called system engineering.

It is very essential to keep a common knowledge of industrial process to bring a better productivity.

INDIA is third largest country of steel production, here we discussed the industrial process of a integrated steel plant that might be helped for knowledge, to be a good engineer.

The Industrial Training pursued by student/engineer has helped in realizing the work done in an integrated Steel Plant, the practical applications of the various instruments. It has very much helped me in realizing the work of the huge machineries in the industries.

Training also helped in preparing a mindset of mine to work in plants where the need of engineers are more and the perfection and responsibility those they carry.

What are major products of steel industry?

It's has wide range of steel products-both Long & Flat.

Among Long products are: structural, Crane Rails, Bars, Rods & Rebar's, wire rods; and Flat products covering range of HR Coils, Sheets& Skelp, Plates ,CR Coils & Sheets, Tinplates, Electrical Steel etc. It also produces Tubular products and Railway products such as rails, wheels, axles and wheels sets.

Other products of SAIL including Pig iron and Fertilizers such as Calcium Ammonium nitrate, Ammonium Sulphate and Coal Chemicals like Benzene, Toluene, and Xylene etc.

Which are the major customers segments serviced by steel industry?

Besides supplying its full range of products to institutional buyers like Defence and the Railways, it is successfully servicing the requirements of variety customers in the following user segments:

- Projects
- Construction

- Heavy engineering
- Fabrications
- Tube manufacturers
- Cold-reducers
- Auto segments
- Cycle
- Drum & Barrel
- Container
- White goods
- Transportation (oil/gas/water)
- Coated sheets manufacturers
- Wire drawers
- Agricultural equipment

Centre for Human Resource Development

Always attached maximum importance on proper training & development of its employees. Its Centre for Human Resource Development has all modern facilities including state-of-the-art Electrical & Electronics laboratory, Hydraulics and Pneumatics laboratory and workshop for effective training and development of its employees.

INTEGRATED STEEL PLANT

What is steel?

Steel is an alloy of iron; it is a mixture of iron ore, limestone, coke & sinter where the carbon percentage is equal to 2% or less than 2%.

Basic features of steel:

- Steel Materials are very cost effective.
- It is very hard material. In such low cost no other that much of hard material is available.
- Steel has huge applications & benefits in different areas.

Integrated steel Plant

Integrated steel Plant is the steel plant where iron ore as well as semi finish and finish product are produced. Total procedure of producing steel related products is described through a flow chart.

SAFETY MEASURES IN PLANT

Introduction

Human life is very valuable and safety at work sites is of very importance. All individuals doing work are exposed to risks, and workers at industrial facilities are exposed to higher risks. The steel plant, due to very complex nature of production processes, material handling and other related function of iron and steel making, are exposed too much greater risks. In the steel plant cook oven plant, sinter plant & blast furnace are the most hazardous areas.

The study

A study of accident figures of a steel plant showed that:

- About 68% of the accidents happen while doing regular jobs that are common in nature of those similar departments.
- At steel works, the age group of 41-45 contributed more to the number of accidents. It is concluded that age group is more prone to accidents. Special attention should be needed for those age group people.

- Road accidents at steel plant are a major single contributor to total accidents. Furthermore, it appears that road accidents are a result of unsafe behavior of the workers, mainly over speeding and not maintaining the traffic rule.
- About 60% of the accidents are resulted from human error and lack of personal alertness.
- It is also observed that more accidents resulted when a person is alone .When working in a groups colleagues may correct the unsafe behavior of each other.
- Iron making group witnessed the highest number of accidents.

- By analyzing injuries, leg injury was the highest. Thus, the message that leg injury is the most common type, it is important to communicate this fact to all employee and contract workers.

- Unskilled contractor workers are more prone to injury. Almost all contractor accidents happened to unskilled contractor labour.

Prevention

Most of the accidents are due to ignorance of the hidden hazards of the job. The safest way to stop injuries is to stop production which is impossible. But to work safe is the condition of our work and to ensure that some precaution have to taken during cast house operation

- Use of P.P.E (Personal Protective Equipment)
- Good physical fitness
- Always be alert and careful
- Knowledge about nature of the job

PLANT OVERVIEW

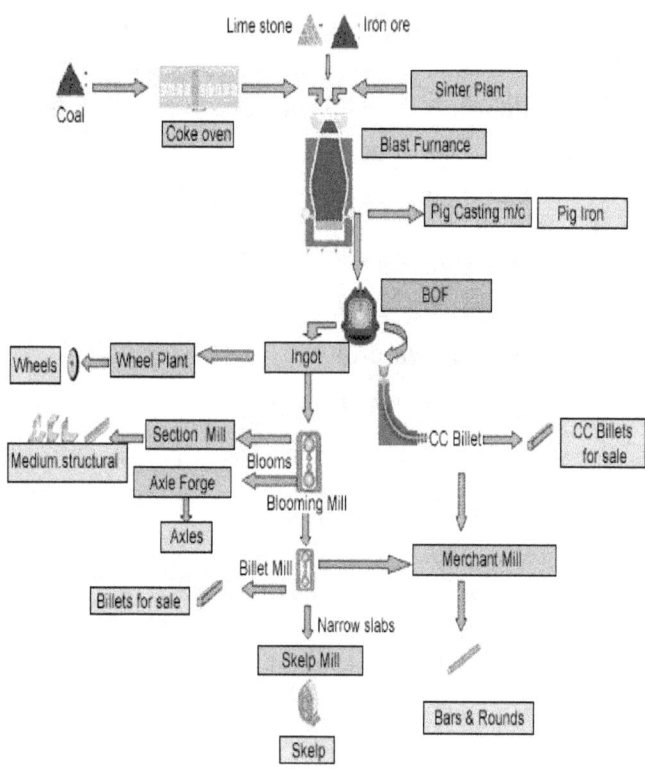

SINTER PLANT

What is Sinter?

Sinter is a hard porous produced by incipient fusion combustion developed within the mass itself. It is agglomerated from the iron ore fines, limestone and dolomite fines, flue dust and coke fines.

Sintering is the agglomeration of fine – grained iron ores for blast furnace burden preparation. Manganese ores can also be sintered before smelting in the electrical arc furnace. Sintering produces a feed of extremely consistent quality in terms of it's:

- Chemical composition
- Grain size distribution
- Reproducibility
- Sinter strength

Sinter plant plays a vital role in the modern steel making industries. It has following advantages:

1. It has sufficient strength, porous giving more surface area for reaction, partial (incipient) fusion leaves iron ores.
2. Utilization of all iron ore fines which have increased due to mechanization of iron ore mining, 40% of iron ore fines is produced during mining.

3. Utilization of all waste materials like coke breeze, mill scale, carbide sludge, lime dust, flue dust.

Sintering process

A Sinter Plant typically comprise of the following sub-units as shown below.

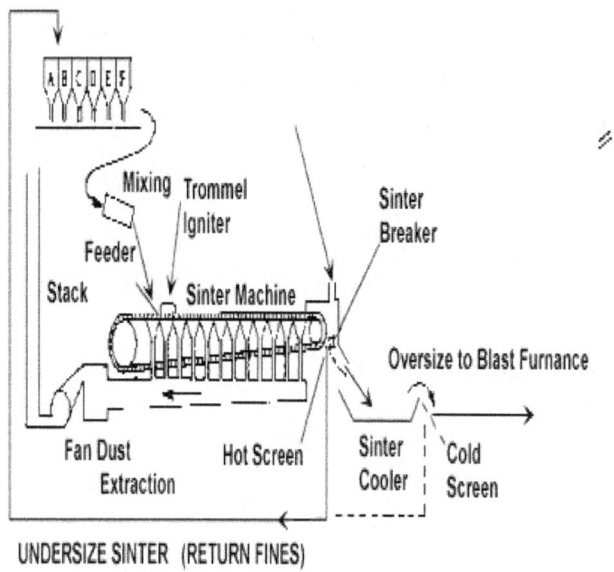

The raw materials are used as follows – iron ore fines (10mm)-75%, coke breeze (3 mm)-3%, Lime stone & dolomite fines (3 mm)-22% and other metallurgical wastes. Coke has same function as the salt in a food, if the quality of the sinter is not desired, extra coke can be added later. The proportioned raw materials are mixed and moistened in a mixing drum. Here infrared absorption type moisture measurement is done here. Ultrasonic level measurement sensors are used to measure the level of the material in the hopper.

The mix is loaded in a sinter machine through a feeder onto a moving grate (pallet) and then the mix is rolled through segregation plate so that the coarse materials settle at the bottom and fines onto the top. The top surface of the mix is ignited through stationary burners at 1200°C.As the pallet moves forward, the air is sucked through wind box situated under the grate. A high temperature combustion zone is created in the charge bed due to the combustion of solid fuel of the mix and regeneration of heat of incandescent sinter and outgoing gases. Due to the forward movement of pallet, the sintering process travels vertically down. The different zone created on a sinter bed is shown in the adjoining figure.

Sinter is produced as a combined result of locally limited melting, grain, boundary and recrystallisation of iron oxides.

On the completion of sintering process, finished sinter cake is crushed and cooled. The cooled sinter is screened and greater than 25 mm & (16-25) mm fractions are dispatched to blast furnace and (6-16) mm & less than 6 mm fractions are recirculated as a return sinter.

Here total 10 bunkers are used; materials are dispatched through conveyor belts. Entire bunkers & conveyor belts movement are controlled through PLC & survisory weight control system.

RAW MATERIAL HANDLING PLANT

Raw Materials

Iron ore, coal and limestone are the three basic raw materials for the steel industry. Steel Plant draws its coal from the adjacent coal belt. A good amount of prime coking coal, having fairly low ash content, is also imported. Bulk of iron ore lumps and fines come from the mines. Lime stone comes from a variety of sources: Birmitrapur (Orissa), Jaisalmer (Rajasthan), and Jukehi and Nandwara (Madhya Pradesh). Other materials used are:

- Return sinter which comes from sinter plant itself. These are the undersized sinters which cannot be supplied to the Blast Furnace.
- BOF(Basic oxygen furnace) slag basically Cao,etc.

Raw Materials Handling

To improve and ensure consistency in raw material quality, the facilities, which have been installed, are:

- Beneficiation/washing facilities, both for lump ore and fines at mines.
- Screening of lump iron ore inside the plant,
- Selective crushing of coal at Coal Handling Plant,
- Base blending facilities for Sinter Plant,
- Silo-cum- Blending bunkers

As part of the modernisation programme, new raw material handling storage and blending facilities with selective crushing of coal have been installed in order to ensure consistency in raw material quality.

Blend mix

In this plant above mentioned different raw materials are mixed together according to the percentage (ore fine 66%, limestone 8%, dolomite 8%, Coke fine 4%, Return sinter 12%, BOF slag 2% etc),the mixer is called blend mix.

Different processes

In this plant different processes are done like

- Crushing
- Blending(mixing)
- Screening(sizing)

Ore fine & return sinter are directly sent to the bunker; limestone, coke & dolomites are crushed first, and then send to the bunker, in the case of lime ore screening is done first then sent to blast furnace.

Here total 3 wagon tipplers, 1 primary & 4 secondary blender recliners, 1 primary & 3 secondary stacker, 3 primaries & 3 secondary crushers, 14 weight feeders, 200 conveyor belts, 4 S.C.R (stacker cum recliner) are used.

In the digital input-output system is used in stacker (4 i/p & 4o/p) & in the case of power & control cabling (8 i/p & 8o/p). Here for wireless transmission purpose Lotus Wireless is used.

Customers of Raw material handling plant

- Sinter plant
- Blast furnace
- Basic oxygen furnace
- NLCP (New line calcinations plant)

Different dispatchers

Total 5 dispatchers are used here:

- Dispatcher 1 is used screening & wagon tippling
- Dispatcher 2 is used bedding & coke crushing
- Dispatcher 3 is used flux – bedding
- Dispatcher 4 is used flux crushing
- Dispatcher 5 sends material to sinter plant & basic oxygen furnace.

All the dispatchers' action is controlled through a C.C.R (central control room). Control actions are done with the help of PLC system of Siemens. A picture of raw material handling plant is given here.

Steel Plant consumes about 7.4 million tonnes of different raw materials annually which comprises over 1.84 million tonnes of coal and 2.9 million tonnes of iron ore lump and fines. Besides the two major raw materials, the plant also requires limestone, dolomite, manganese ore, bauxite, silico manganese, ferro manganese, ferro silicon, etc.

COKE OVEN PLANT

Coke is used in Blast Furnace (BF) both as a reductant and as a source of thermal energy. It involves reduction of ore to liquid metal in the blast furnace and refining in converter to form steel.

Coke Making - Coal carbonization

Coking Coals are heated in the absence of air, first melt, go in the plastic state, swell and resolidify to produce a solid mass called coke. When Coking Coals are heated in absence of air, a series of physical and chemical changes takes place with evolution of gases and vapours, it helps to produce coke.

In this plant, three kinds of coals are mixed together and make a blend. Three kinds of coals are:

- PCC (Primary coking coal)
- MCC (Medium coking coal)
- Imported coal

PCC & MCC are local coals – (24-25 % ash), this type of coals are brought from Raniganj, Charmela, Raypura etc area & the imported coal-(8% ash), are brought from New Zealand, USA, Australia, Canada, China etc country.

14% PCC, 12% MCC, 74% imported coal are mixed to produce a blend mix. Conventional coke making is done in coke oven battery where ovens are sandwiched between heating walls. Coals are carbonized at a temperature around $1000°C - 1230\ °C$ up to a certain degree of devolatization to produce metallurgical coke of desired mechanical and thermo-chemical properties.

At first crushing coals are put on the charging car, then from the charging car the coals are charged inside the battery through 4 holes, situated upper portion of the battery.

Then there is another car outside of battery called ramp car which consists of two basic parts – leveller & pusher. The function of leveller is to distribute equally the coals charging through the upper holes. Pusher pushes the coke outside of the battery after completing the process. Here two gases (Blast furnace gas, Coke oven gas) are sent to the battery from lower part of the battery, gases are helped in carbonization process. Gas flow is maintained with help of butterfly valve.

During carbonization process coking coals are undergo transformation into plastic state at around $(350-450)\ °C$, swell and then resolidify at around $(500\ -559)\ °C$ to give semi-coke then coke. In coke ovens, after coal is charged inside the oven, plastic layers are formed adjacent of the heating wall. And with the progress of time, the plastic layers are move towards the centre of the oven from either side and ultimately meet together at the centre. In Coke oven Battery the cooking time is (19-20) hours. After completing the process cokes are cooled then send to blast furnace through conveyor belt. A schematic diagram of Coke oven Battery is given in Fig below.

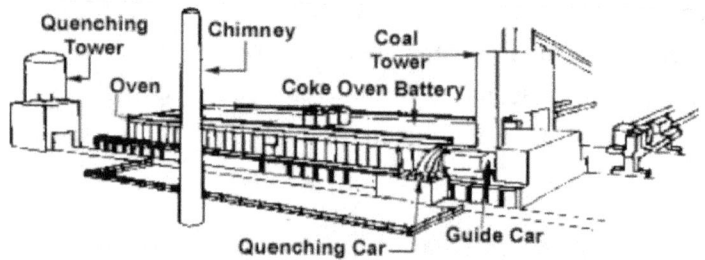

In Steel Plant consists of 6 batteries. Each battery are divided into two group A & B. Each group consists of 39 ovens. Among six batteries, battery 2 is newly reformed & battery 5 is sold out. Now total no of assembly of ovens is 273.

Here three kinds of cokes are produced:

- 20 mm send to blast furnace
- (10-20) mm & (0-10) mm send to raw material department.

BLAST FURNACE

The purpose of blast furnace is to chemically reduce and physically convert iron oxides into liquid iron called "hot metal". The blast furnace is a huge, steel stack lined with refractory brick, where iron ore, coke & limestone are dumped into the top, and preheated air is blown into the bottom. The raw materials require 6 to 8 hours to descend to the bottom of the furnace where they become the final product of liquid slag and liquid iron. These liquid products are drained from the furnace at regular intervals. The hot air that was blown into the bottom of the furnace ascends to the top in 6 to 8 seconds after going through numerous chemical reactions. Once a blast furnace is started it will continuously run for 4 to 10 years with only short stop of perform planned maintenance.

The process

The iron oxides come to the furnace in the form of iron ore. The raw ore is sized into pieces that range from 0.5 to 1.5 inches. This ore is either Hematite (Fe_2O_3) or Magnetite (Fe_3O_4) and the iron content ranges from 50% to 70%.

Sinter is produced from fine raw ore, small coke, sand-sized limestone and other wastage materials that contain some iron.

These iron ore and sinter are charged into the furnace, produce liquid iron with other remaining impurities going to the liquid slag.

The coke, which is produced from the cook oven plant, contains carbon (90 – 93) %, some ash and sulfur. The strong pieces of coke with a high energy value provide permeability, heat and gases which are required to reduce and melt iron ore, sinter.

The final raw material in the steel making process is limestone. It is first crushed and screened to the size that ranges from 0.5 to 1.5 inch to become blast furnace flux. The flux can be pure high calcium limestone, dolomite limestone containing magnesia or a blend of two type of limestone.

Since limestone is melted to become a slag which removes sulfur and other impurities, the blast furnace operator may blend the different stones to produce the desired slag chemistry and create optimum slag properties such as a low melting point and a high fluidity.

Once the materials are charged into the furnace, they go through several chemical reactions. At first oxygen in the iron oxides is removed by a series of chemical reactions:

$3Fe_2O_3 + CO = CO_2 + 2 Fe_3O_4$ Begins at 850°F

$Fe_3O_4 + CO = CO_2 + FeO$ Begins at 1100°F

$FeO + CO = CO_2 + Fe$ Begins at 1300°F

$FeO + C = CO + Fe$ Begins at 1300°F

At the same time iron ore, sinter is beginning to soften then melt and finally trickle as liquid iron through the coke to the bottom of the furnace. The coke descends to the bottom of the level where the preheated air or hot blast enters the blast furnace. The coke is ignited by this hot blast immediate reacts to generate heat as follows:

$C + O_2 = CO_2 + Heat$

At high temperature carbon dioxide is reduced to carbon monoxide which is necessary to reduce the iron ore.

$CO_2 + C = 2 CO$

The limestone descends in the furnace and remains a solid while going through its first reaction:

$CaCO_3 = CaO + CO_2$

This CaO is used to remove sulfur from the iron which is necessary before the hot metal becomes steel.

$FeS + CaO + C = CaS + FeO + CO$

CaS becomes part of the slag. The slag is also formed from any remaining Silica (SiO_2), Alumina (Al_2O_3), Magnesia (MgO) or Calcia (CaO) that entered with iron ore, sinter or coke. The liquid slag when trickles through the cook bed to the bottom of the furnace where it floats on top of the liquid iron since it is less dense.

Another product of the steel making process, in addition to molten metal and slag, is hot dirty gas. These gases exit top of the blast furnace and processed through gas cleaning equipment where particulate matter is removed from the gas and the gas is cooled. This gas has a considerable energy value so it is used as a fuel in the "hot blast stoves" which are used to preheat the air entering the blast furnace to become "hot blast ".

Here K-Type thermocouple is used for temperature measurement. Very high temperature measurement (like- hot metal) purpose radiation pyrometer is used. Total blast furnace working is control through a control room.

Typical hot metal chemistry follows:

Iron (Fe) = (93.5 – 95) %

Silicon (Si) = (0.30 – 0.90) %

Sulfur(S) = (0.025 – 0.050) %

Manganese (Mn) = (0.55 – 0.75) %

Phosphorus (P) = (0.03 – 0.09) %

Titanium (Ti) = (0.02 – 0.06) %

Carbon (C) = (4.1 – 4.4) %

Durgapur steel plant consists of 4 blast furnaces (BF) – BF1, BF2, BF3, and BF4. Among them BF1 is now not working. Basic comparison of BF2, BF3, and BF4 are given here:

Feature	BF2	BF3	BF4
Rated capacity (tones/day)	1820	1820	2340
Useful volume (m^3)	1400	1400	1800
Top charging equipment	Double bell system	Double bell system	Double bell system
Hot stoves (to heat ambient air)	3 dedicated stoves	3 dedicated stoves	3 dedicated stoves
Slag granulation system	Not use	Use	Use

A picture of blast furnaces of steel plant is given here.

STEEL MELTING SHOP

Steel melting shop is a very vital part of a steel plant. Here liquid iron converts into liquid steel. The whole operation takes place in a converter, Durgapur steel plant consists of total 3 numbers of converters, and While 2 converters are working another one is kept stand by. Capacity of each converter is about (110-130) tonnes. Each converter is driven by 2 motors, individual capacity of 100 kilo-watts. Inner surface of converter is made of Tar bound Cr-Mg bricks.

At first scrap (10-15) tonnes charge in the converter, after that hot metal (110-115) tonnes charge in this converter. Thereafter oxygen is blown in the mixture through a lance pipe. The lance pipe is entered in the converter from upper direction. The oxygen is blown about (16 – 17) min with a velocity of 415 m³/min. Then, exothermic reactions start inside the converter. Then the temperature increase up to 1700°C. The reactions are given below:

$$FeO + Si \longrightarrow SiO_2 + Fe$$

$$FeO + Mn \longrightarrow Mn + Fe$$

$$FeO + P \longrightarrow P_2O_5 + Fe$$

$$SiO_2 + CaO \longrightarrow Ca_3(SiO_4)_2$$

At last N_2 splashing is done in converter to protect the refractory bricks. After the process over two layers are produced in the converter – liquid steel & slag. Then liquid steel is brought outside of the converter through a tap hole & put in a ladle, send it to continuous casting plant.

A picture of steel melting shop of Durgapur steel plant is given here:

In this plant quality of the steel is maintained. Here the percentage of carbon, sulfur, phosphorus, silicon, manganese is maintained at a certain level according the quality of the steel.

S-type thermocouples, radiation pyrometers are used here for used here for high temperature measurement & for low temperature measurement Pt-100 type RTDs are used. Orifice measures the flow of oxygen gas; magnetic flowmeter measures the flow of water line.

Whole temperature, pressure & flow control system is controlled by the DCS system.

Continuous casting plant

The state of the art CCP has 2 machines having 6 strands each. The other basic details are as follows: -

Design limits- 80-150 sq .mm, casting radius- 6 metres
Casting time – 85 minutes, Cut-off lengths- 6 / 9 / 12 metre
No of ladle treatment stations-2
Mould level controller - Automatic (Radio-active Co-60)

The steel ladle from BOF is taken to the ladle treatment station. At the ladle treatment station, liquid steel is rinsed with nitrogen to homogenise its temperature and composition. After the rinsing, the ladle containing liquid steel is placed on the turret and brought over the tundish. The tundish acts as a buffer and enables the liquid steel to move homogeneously down through the six nozzles, provided at the bottom of the tundish into moulds. The automatic mould level controller controls the steel level in the mould. The subsequent primary and secondary cooling transforms the liquid steel into billets of the required dimensions and is drawn out with the help of a withdrawal and straightener unit and cut into the required length by the shear provided in each strand. The continuous casting process is the result of a unique synchronisation between Basic Oxygen Furnace and CCP. Once a ladle is emptied, another ladle is brought into casting position and the casting continues. The billets are gradually shifted to the cooling beds and then stacked orderly at the despatch end for outside despatch. The details about the cast number and quality of the billets are marked on the billet stack. The Merchant Mill of Durgapur Steel Plant utilises billets for rolling TMT bars and other merchant rounds, while a sizeable portion is sold in the domestic and foreign markets.

ROLLING MILLS

Ingots weighing 8 tonnes each are heated in the soaking pits (numbering 20) for about 7 to 12 hours at around 1,200 degrees centigrade and thereafter rolled in the 42" primary and the 32" secondary blooming mills. These are rolled further into different shapes and sizes in different finishing mills.

BLOOMING MILL

Installed Mill capacity - 1.47 million tonnes/year
Ingot weight - 8 tonnes
42" Mill:
42" x 102" reversible Blooming Mill
Output bloom size (min) - 300 mm x 250 mm

32" Mill:
32" x 84" reversible Intermediate Mill
Output bloom size (min) - 180 mm x 180 mm

BILLET MILL

Installed Mill capacity - 0.957 tonnes / yr.
Type - Continuous Morgan design
Horizontal stands - 6, Vertical stands - 2

Product Range

Billets	- 100 mm square to 125 mm square
Sleeper bars	- 352 mm x 12.5 mm
Skelp slabs	- 140 mm x 75 mm to 240 mm x 90 mm

The ingots after heating are rolled in the Blooming Mill to make blooms of the sizes mentioned in the table and then a part of the same are then further rolled in the Billet Mill for making rolled billets or slabs as per the above details.

SECTION MILL

The Section Mill rolls out light and medium structural like joists, channels and angles.

Mill capacity	- 0.2 million tonnes / year
Re-heating furnaces	- 2 x 40 t/hr
Roughing Mill	- 2 high reversible
Intermediate Mill	- 2 stands of 3 high non-reversible
Finishing Stand	- 2 high non-reversible

Product range:

Joists   - 200 mm x 100 mm, 175 mm x 85 mm
             150 mm x 75 mm, 116 mm x 100 mm
Channels - 200 mm x 75 mm, 175 mm x 75 mm
             150 mm x 75 mm, 125 mm x 65 mm
Angels   - 150 x 150 mm, 130 x 130 mm
             110 x 110 mm, 100 x 100 mm

Fish plate bars for 52 kg rails

MERCHANT MILL

The Merchant Mill produces plain round and Thermo-Mechanically Treated (TMT) bars in the range of 16mm - 28mm. The entire product range of TMT bars and rods at DSP is branded and has been able to create a niche market.

Plain round and TMT bars are produced from Billet (100mm × 100mm × 9 meters). This Billets are brought from Billet Mill.

Billet ⟹ | MARCHENT MILL | ⟹ Round & TMT Bars

Generally produces three kinds of TMT bars according to diameter:

10 mm, 16mm, and 25mm.

Plain round bar diameter: (12-32) mm

To produces this kind of bars following procedure are needed:

- Large tundish volume for inclusion floating
- Steel flow control
- Hydraulic mould oscillation
- Tubular mould
- Eddy current mould level measuring
- Electromagnetic mould, final stirrer
- Cooling using rapid water quenching system
- Dynamic mechanical soft reduction process must for the ultra soundness of the big cross section.

Etc.

At first give heat to the billet material, then rolling it: here at a time 160 billets can be rolled with rolling rate 70 tonnes /hour. Cutting it according to shapes are required, straightening, packing, loading and dispatch it.

In the Merchant Mill of DSP total 12 motors are used:

- 4 motors capacity: 400HP
- 3 motors capacity: 500HP
- 3 motors capacity: 600HP
- 2 motors capacity: 800HP

Capacity   - 0.28 million tonnes / year
Type of mill   - continuous Morgan design
Horizontal stands - 13, Repeaters - 4

Product range

Plain rounds   - 12 - 32 mm dia
TMT bars   - 12 - 25 mm dia

ELECTRICAL TECHNICAL LABORATORY

Electrical technical laboratory is one of the important laboratories in DSP. Here different kinds of instruments, systems are come for repair purpose from different plant of DSP. Lists of different kinds of instruments, systems are repaired given below:

- **Weight measuring system**

Here two kinds of weight measuring systems are brought for repairing:

> - System which can measure the static loads. Like: at the time of delivery of material static weight measurement is done.

> - System which can measure the loads in motion. Suppose there is a need to measure the weight of a loaded wagon at the time this type of weight measuring system is required.

- PLC systems, which are very essential for controlling purpose, are also repaired here.

- Thyristor, which is a four layered semiconductor rectifier, can regulate voltage changing its thyring angle. These types of devices are also rectified here.

- Monitors, peripherals, hardware are also repaired here, etc.

INSTRUMENTATION WORKSHOP

Instrumentation department of Durgapur steel plant have seven units, among them central Instrumentation workshop is important one. Different electromechanical, electro pneumatic, electronics instruments are brought here to calibration & rectification purpose from different units.

Here two types of calibration are done: normal calibration & ISO calibration.

Lists of instruments are given below:

- S – type, R– type, K– type thermocouples are manufactured here. Also radiation pyrometers, Pt-100, Ni-100 type RTD are rectified and calibrated here.
- Capacitive level sensors, ultrasonic level sensors are also calibrated here.
- Different types of pressure gauge are calibrated here with the help of dead weight tester.
- Smart transmitters are rectified in this workshop.
- Micro manometer, U-type manometer are calibrated here.
- Positioners, control valves are repaired here.
- Scanners, magnetic flowmeters, analyzers are also repaired here, etc.

WHEEL AND AXLE PLANT

The first consignment of 143 EMU wheels made in Steel Authority of India Limited's Durgapur Steel Plant's Wheel & Axle Plant.

Durgapur Steel Plant is the only major indigenous supplier of wheel sets, loco wheels, carriage and wagon wheels, and axles to the Indian Railways. As per demand of the Railways, the plant has developed loco wheels, which were imported earlier. The Wheel & Axle Plant is producing wheels manufactured as per the latest IRS specifications, i.e. R-19/93 for carriage and wagon wheels, R-34/99 for loco wheels and R-16/95 for axles. They produce it in collaboration with KLW of Switzerland.

The wheel plant of the Wheel & Axle Plant is provided with six PLC controlled band saws for accurate slicing of the 14" and 16" fluted ingots. A fully computerised 63/12 MN oil hydraulic press is there for forging and punching of the wheel blanks along with a fully computerised vertical wheel mill and other downstream facilities. All the wheels are 100 per cent rim-quenched, tempered and tested as per IRS specifications.

Machining of these forged rolled and heat-treated wheel blanks are carried out in the 15 CNC machines. All the wheels are ultrasonically tested and inspected by RITES on behalf of the Indian Railways. A number of sophisticated and modern online testing facilities are there to conform to the stringent testing requirements of the Indian Railways.

Steel production of the railways wheels and tyres is melted in the open hearth furnaces with the carriage of 250 tonnes and tapping in two ladles. High quality characteristics of steel are provided by subsequent processing at the out-of-furnace complex of steel treatment. The ladles with the metal are delivered to unit furnace-ladle where metal finishing and refining takes place. Steel blowing in the ladle by argon alone with the refining process provide for the sulfur content in the finished metal equal to 0.010% ,phosphorus 0.015% or less, and uniform distribution of another elements.

Degassing process means removing of hydrogen, nitrogen, oxygen from the metal. It is done simultaneously with blowing argon. Hydrogen content in the steel is limit up to 2 ppm.

At last weight of each wheel, distance between two wheels in axle is kept constant. Hardness of the wheels are kept (260-290)BHN for coaching wheels, (300-340)BHN for loco wheels.

Production Rate

Annual production of finished wheels — 1,00,000 nos.
Production rate in rolling/forging — 25 nos./hr
Production rate in machining — 22 nos./hr

Above diagram shows the production of wheel in Wheel & Axle Plant.

Engineering shops

Steel Plant has a number of captive engineering shops for repairs and supply of spare parts. The Central Engineering Maintenance has a Machine Shop, Structural Shop, Fitting and Assembly Shops. The Foundry produces Ingot moulds and bottom plates for the steel melting shop. There are also Auxiliary Repair Shops such as Electrical, Wagon and Loco repair.

Research and Control laboratories

The Research & Control laboratories are entrusted with the responsibility of maintaining quality of products and also developing new products. It is well equipped for carrying out sophisticated chemical, metallurgical and other tests.

Computerisation

An extensive computerisation has been undertaken in DSP for personnel, commercial, process control, and production and maintenance applications. The Production Planning and Control network is thoroughly used for tracking of customer orders, material, monitoring of quality parameters and ensuring availability of accurate, real time data to all agencies needing access to the data.

Quality Assurance

In order to be fully competitive on the quality front, Durgapur Steel Plant has set out to acquire ISO 9000 certification for all its units. The Merchant Mill is the first to secure the prestigious ISO 9002 certificate. Subsequently, steel melting shop, basic oxygen furnace shop, continuous casting plant, and wheel and axle plant were also awarded the ISO 9002 certification and recently the Skelp Mill has been awarded the ISO 9002 certification.

CHAPTER-2

PROCESS OF MACHINING

INTRODUCTION

Machine Tools, stationary power-driven machines used to shape or form solid materials, especially metals. The shaping is accomplished by removing material from a workpiece or by pressing it into the desired shape. Machine tools form the basis of modern industry and are used either directly or indirectly in the manufacture of machine and tool parts.

Machine tools may be classified under three main categories: conventional chip-making machine tools, presses, and non-conventional machine tools. Conventional chip-making tools shape the workpiece by cutting away the unwanted portion in the form of chips. Presses employ a number of different shaping processes, including shearing, pressing, or drawing (elongating). Non-conventional machine tools employ light, electrical, chemical, and sonic energy; superheated gases; and high-energy particle beams to shape the exotic materials and alloys that have been developed to meet the needs of modern technology.

Non-conventional machine tools include plasma-arc, laser-beam, electro discharge, electrochemical, ultrasonic, and electron-beam machines. These machine tools were developed primarily to shape the ultra hard alloys used in heavy industry and in aerospace applications and to shape and etch the ultrathin materials used in such electronic devices as microprocessors.

DRILLING MACHINE

Hole-making machine tools are used to drill a hole where none previously existed; to alter a hole in accordance with some specification (by boring or reaming to enlarge it, or by tapping to cut threads for a screw); or to lap or hone a hole to create an accurate size or a smooth finish.

Drilling machines vary in size and function, ranging from portable drills to radial drilling machines, multispindle units, automatic production machines, and deep-hole-drilling machines.

Boring is a process that enlarges holes previously drilled, usually with a rotating single-point cutter held on a boring bar and fed against a stationary work piece. Boring machines include jig borers and vertical and horizontal boring mills.

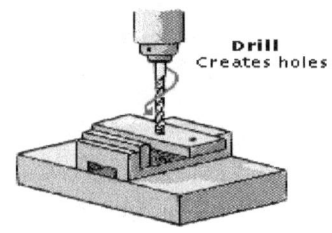

OPERATION OF DRILLING MACHIINE

1. DRILL FOR MAKING A HOLE
2. BORING
3. TAPPING
4. REAMING
5. PARTING (CHAIN DRILL)
6. COUNTERSINKING
7. COUNTER BORING
8. SPOT FACING
9. LAPPING
10. GRINDING
11. TREPANNINGS

RADIAL DRILL (RM62)

SPECIFICATION FOR RADIAL DRILLING MACHINE

MAIN MACHINE	UNIT	RM62
Capacity		50
Drilling in solid, steel 50 kgf/mm² cl 180-200 BHN	mm	60
Taping ,metric fine steel cl	mm	56×2.5 70×2
Drilling radius max. min.	mm	1500 530
Sleeve dia	mm	350
Distance spindle to distance max. min.	mm	1325 355
Spindle traverse	mm	325
Distance, Base to spindle max. min.	mm	1420 355
Working surface of base length width	mm	1490 910
Standard box table length×width×height	mm	600×500×500
Drilling spindle Taper in spindle		MT5 12

| nose Spindle speed | No. | |

FEEDS

Feeds range	mm/revolution	0.125 to 1.25
Max. drilling thrust	daN	1650

POWER

Drill head motor	kw	3.6/4.5
Elevating motor	Kw	1.50
Coolant pump motor	Kw	0.11
Hydraulic pump motor(EHCL)only	Kw	0.37
Machine weight approx.	Kg	3160
Shipping dimensions length width height	mm	2.81 1.28 2.91

GRINDING MACHINE

Grinding is the removal of metal by a rotating abrasive wheel; the action is similar to that of a milling cutter. The wheel is composed of many small grains of abrasive, bonded together, with each grain acting as a miniature cutting tool. The process produces extremely smooth and accurate finishes. Because only a small amount of material is removed at each pass of the wheel, grinding machines require fine wheel regulation. The pressure of the wheel against the work piece can be made very slight, so that grinding can be carried out on fragile materials that cannot be machined by other conventional devices.

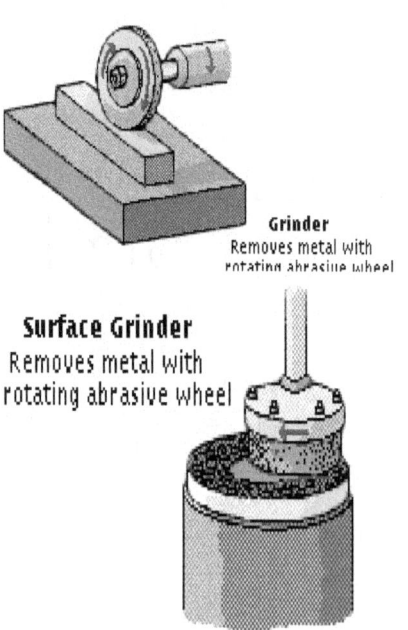

Grinder
Removes metal with rotating abrasive wheel

Surface Grinder
Removes metal with rotating abrasive wheel

- OPERATION OF GRINDING MACHINE
    1. SURFACE FINISH
    2. KEY SLOT
    3. ACCURATE THICKNESS
    4. SPECIFIC PROFILE
    5. THREAD
    6. GEAR TEETH
    7. FILLET
    8. FORMED SURFACE
    9. TAPERED SURFACE

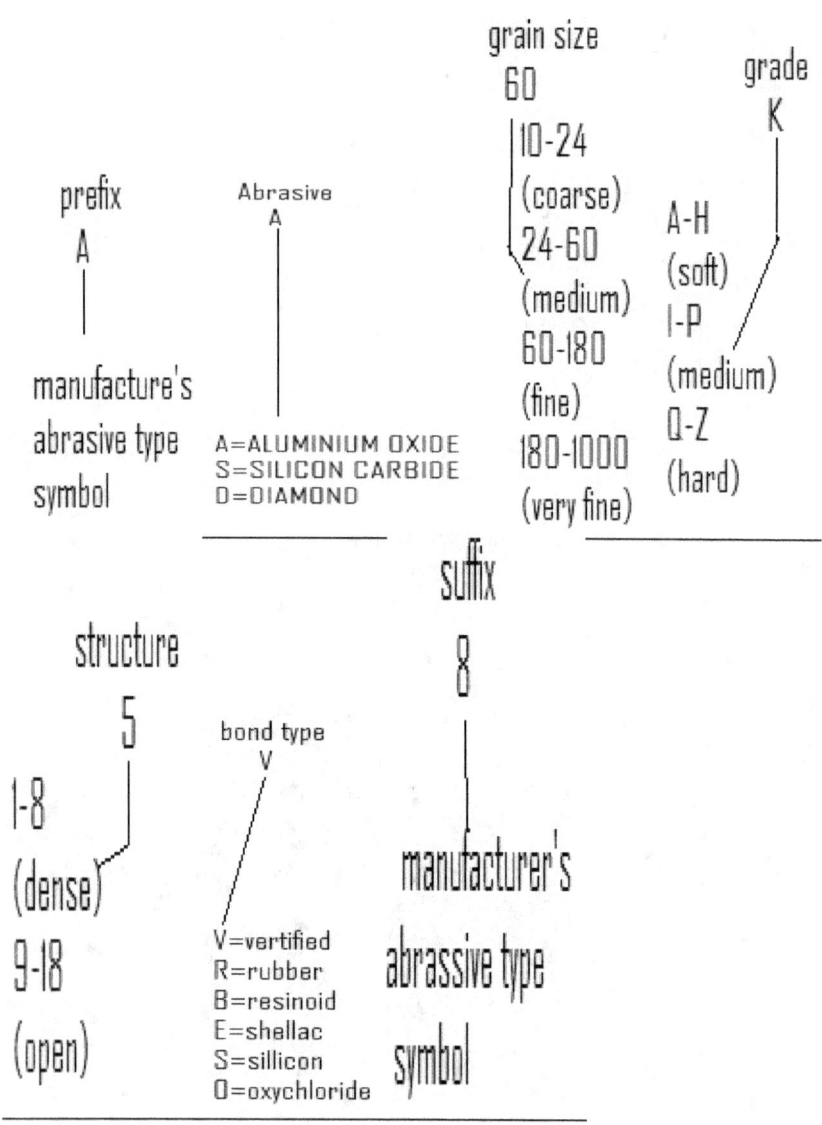

INDIAN STANDARD MARKING SYSTEM

- SPECIFICATION OF SURFACE GRINDING MACHINE

MAIN MACHINE	UNIT	SFW 250×1000
Working surface	mm	250×1000
Number of T-slots		1
Max. transverse, traverse of grinding wheel	mm	320
Max. long, traverse of table	mm	1150
Max. width to be ground with relief at the front & the rear	mm	250
Max. width to be ground without relief at the front & the rear	mm	380
Max. grinding length	mm	1000
Max. grinding height with new wheel	mm	400
Infinity variable table speeds	m/mm	2.5-25
Automatic continuous traverse feeds of grinding wheel	m/mm	0.3-3.5
Automatic intermittent traverse feed of grinding wheel per table stroke	m	1-32
Precision traverse adjustment of grinding wheel by handle graduation	mm	0.002

Rapid vertical speed of grinding wheel	m/mm	0.6
Grinding wheel speed approx.	RPM	1440/2880
Diameter upto	mm	300
Width upto	mm	63
Bore upto	mm	76
HP of hydraulic pump motor	HP	3
Speed of hydraulic pump motor	RPM	1440
HP of grinding motor	HP	4/5.7
Speed of grinding motor	RPM	1440/2880
HP of vertical adjustment motor	RPM	0.25
Speed of vertical adjustment motor	RPM	2880
Connect load of the machine(approx.)	HP	11
Max. weight of the job that can be mounted 1.without chuck 2.with chuck	kg kg	150 325
Net weight of the machine without extra equipment	kg	3700
Net weight of the machine when packed with extra equipment	kg	4300
Span require(length×width×height)	m	3.3×1.6×2.3

LATHE MACHINE

Lathe, machine tool that shapes metal, wood, or other material by rapidly turning it against a stationary cutting device. The material to be shaped on a lathe is called the *workpiece*.

Lathes are one of the oldest and most important of machine tools. Lathes can shape, drill, bore, grind, and perform other operations. Woodworking lathes were used as early as the Middle Ages (from about the 5th century to the 15th century). These lathes were usually powered by using a *treadle,* a foot-powered lever that, when pushed down, drove a mechanism that turned the lathe. By the 16th century lathes were powered continuously by hand cranks or waterpower, and were equipped with a cutting-tool holder that enabled more precise shaping of the workpiece. As the Industrial Revolution began in England during the 17th century, lathes were developed that could shape a metal workpiece. The 18th-century development of the heavy industrial metal-cutting lathe made possible precision manufacture, interchangeability of parts, and mass production.

- FUNCTION OF LATHE:

    1. Thread cutting
    2. Step Turning
    3. Tapping
    4. Knurling

5. Boarig
6. Champering
7. Facing
8. Reaming
9. Tapping
10. Parting off
11. Grooving

SPECIFICATION OF LATHE:
1. Maximum diameter of job can be mounted on the lathe
2. Length of the bed
3. Distance between left centre & dead centre
4. Height of the spindle from centre to the bed
5. Maximum RPM of lathe

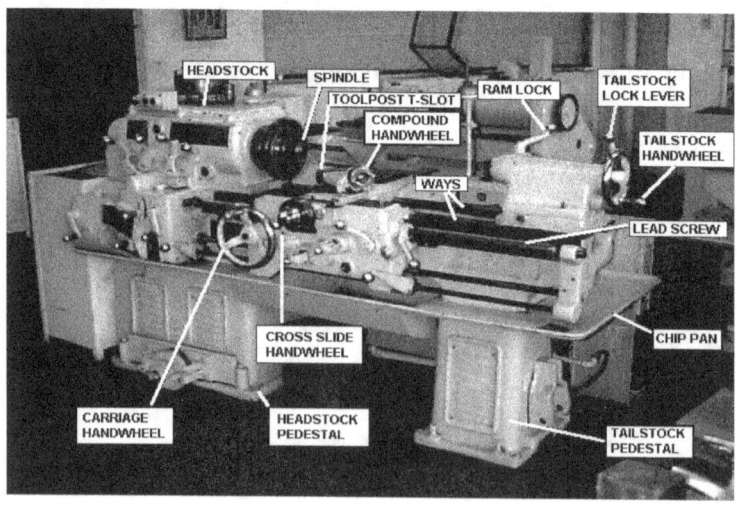

CNC EDM

EDM is known as Electro Discharge Machining. In this process electrical energy is used directly to cut the material to final shape & size. Another advantage of this process is no complicated features are needed for holding the job & even very thin job can be machined to the desired dimension & shape. All the operation are carried out in a single set-up.

The main principle of this process is to ionize the material to be removed. Here tool is the electrode.

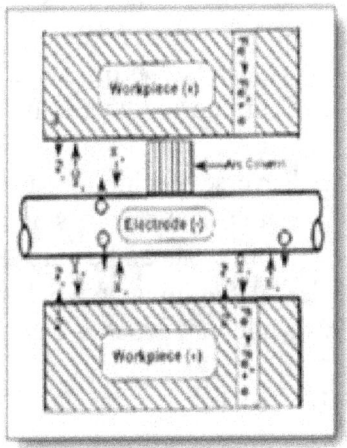

Electric Current Diagram

Electro discharge machining (EDM), also known as spark erosion, employs electrical energy to remove metal from the work piece without touching it. A pulsating high-frequency electric current is applied between the tool point and the work piece, causing sparks to jump the gap and vaporize small areas of the work piece. Because no cutting forces are involved, light, delicate operations can be performed on thin workpiece. EDM can produce shapes unobtainable by any conventional machining process. But despite all that the given process of EDM has some limitation & drawbacks prominent among which is the slowness, expensiveness and inconvenience in achieving the accuracy. That's why EDM is used as the last resort machining process where all other machining procedures fail.

- **ADVANTAGE OF EDM**
    1. Specific surface finish can be obtained
    2. Small rib can be made
    3. Any critical profile can be made
    4. Sharp corner, threading & knurling can be done
    5. Reverse machining can be done
    6. Lettering & embossing can be made
    7. Thin area can be machined
    8. The clamping area is less
    9. Undercut, grooves can be easily achieved

- > EQUIPMENT
    The basic units constituting an EDM equipment are
    - Machine tool structure with work positioning unit.
    - Servo head and tool feed.
    - Power supply.
    - Dielectric fluid-system.

The machine tool structure will depend on the work-tool configuration to perform the required activities. The table below shows the three main categories into which spark-machining operations can be divided:
- Die sinking by EDM.
- Cutting by EDM.
- Grinding by EDM.

EDM WIRE-CUT

The wire cut EDM uses a very thin wire of size from 0.02 to 0.03 mm in diameter as an electrode and machine and a work piece with electric discharge like a band saw by moving either wire or work piece. Erosion of metal utilizing the phenomenon of spark discharge is the very same as conventional EDM. Wire-cut EDM machine basically consists of a machine proper composed of a work piece contour movement control unit.

(NC unit or copying unit), work piece mounting table and a wire drive section for accurate movement of wire at constant tension; a machining power supply which applies electrical energy to the wire electrode; and a unit which supply a dielectric fluid (distilled water) with constant specific resistance.

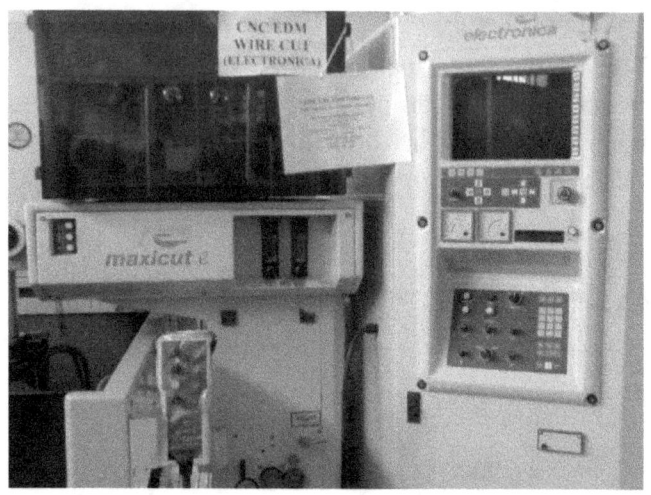

➤ SPECIFICATION OF WIRE-CUT EDM

Travel range:		734 Machine
Longitudinal	Y axis	400mm
	V axis	± 40 mm
Lateral	X axis	300 mm
	U axis	± 40 mm
Vertical	Z axis	225 mm
Work piece size: Table size (WxD) Max. work piece size **Max. Work piece weight**		110×450 mm 400×500×600 mm 400 kg

Feed: Main table feed rate Resolution Wire feed rate Wire tension	170mm/min(900mm/min) .001mm(.0005mm) 0-10m/min(0-15m/min) 1.5kgf(2.5 kgf)
Wire guide: Wire guide type Wire electrode diameter	Diamond close type 0.25 STD 0.15, 0.2, 0.3 OPT
Taper cutting: Max. taper angle	$\pm 15^0/100$

Dielectric Unit Dielectric fluid Tank capacity Paper filter Cooling System	De-ionized water 250 Liters 10 µ Single Cartridge 1700 kCal

CNC DIE-SINKING

SPECIFICATION

MAIN MACHINE	UNIT	SP1
Machine external dimention(L×W×H)	mm	1500×1200×1991
Net weight of main machine	kg	1400
Nominal work piece dimention(L×W×H)	mm	740×450×260
Work tank dimentions(L×W×H)	mm	1000×550×360
Work table dimentions(L×W)	mm	500×320
Work tank capacity	L	178
X travel	mm	320
Y travel	mm	250
Z travel	mm	250
Maximum table workload	kg	600
Maximum ram workload	kg	80

DIELECTRIC UNIT

Reservoir capacity	Liters		290
Dielectric pumps flow rate(single pump)	L/min		120
Dielectric filter medium			Paper cartridge
External dimention	Mm		1200×720×700
Weight	Kg		300

GENARATOR

Model		SP50/SP100(optional)
External dimention	Mm	632×533×1714
Weight	Kg	380
Display unit		ColorCRT
Power consumption	KVA	10/18
Max. machining current	A	50/100

DIELECTRIC PUMP

1Pump	1.1KVA

ENVIRONMENTAL REQUIREMENT

Room temperature For guaranteed precision For operation of machine	20°C±3°C 15°C 30°C
Humidity	40% 80%
Sound emission	<80db

PERFORMANCE

Max.maching current	50A/100A
Best surface roughness	Ra≤0.4µm
Lowest electrode wear	≤0.3%

> DRAW BACK OF CNC EDM:

1. It is very time taking method.
2. Achiving accuracy is very less.
3. Machine cost is very high.

MILLING MACHINE

A milling machine is a machine tool used to machine solid materials. Milling machines are often classed in two basic forms, horizontal and vertical, which refers to the orientation of the main spindle. Both types range in size from small, bench-mounted devices to room-sized machines. Unlike a drill press, which holds the workpiece stationary as the drill moves axially to penetrate the material, milling machines also move the workpiece radially against the rotating milling cutter, which cuts on its sides as well as its tip. Workpiece and cutter movement are precisely controlled to less than 0.001 in (0.025 mm), usually by means of precision ground slides and leadscrews or analogous technology. Milling machines may be manually operated, mechanically automated, or digitally automated via computer numerical control (CNC).

A miniature hobbyist mill plainly showing the basic parts of a mill.

A milling machine is often called a **mill** by machinists. The term **miller** also used to be common (19th and early 20th centuries), although it is typically not used today in reference to modern machines.

CHAPTER-3

INTELLIGENT MANUFACTURING SYSTEMS

Abstract

The Global competition and rapidly changing customer requirements are demanding increasing changes in manufacturing environments. Enterprises are required to constantly redesign their products and continuously reconfigure their manufacturing systems. Traditional approaches to manufacturing systems do not fully satisfy this new situation. Many authors have proposed that artificial intelligence will bring the flexibility and efficiency needed by manufacturing systems. This paper is a review of artificial intelligence techniques used in manufacturing systems. The paper first defines the components of simplified intelligent manufacturing systems (IMS), the different Artificial Intelligence (AI) techniques to be considered and then shows how these AI techniques are used for the components of IMS.)

1. INTRODUCTION

The goal of intelligent manufacturing system is akin to any normal manufacturing system satisfying customer needs at the most efficient level for lowest possible cost. The involvement of computers as in the computer-integrated manufacturing has been for more than 20 years; the incorporation of computer technology does not necessarily result in intelligent manufacturing system. It is the introduction of human like decisions making capabilities into the manufacturing system that makes it indeed intelligent. In particular knowledge base systems have dominated the manufacturing landscape of late 1990's CIM was all rage in 1980s, FMS in 1970s, Manufacturing technology will play an important role in future human developments if it is based on new knowledge based applications. The impact of knowledge-based system on manufacturers is already in front and all of us to see and experience. Any organization must make full advantage of the knowledge at its disposal. This goal translates into the effective use of knowledge ranging from design to production and maintenance.

2. CONCEPT OF MANUFACTURING

Manufacturing is defined very broadly as the process by which material, labor, energy, and equipment are brought together to produce a product having a greater value than the sum of the material put in. This can be shown as a system, as indicated figure 1. Here the input is shown as material labor, energy and capital. The capital input provides the equipment and facilities required for combining the material, labor, and energy. Output includes product, but there is always some undesirable output -waste and scrap—which should not be forgotten. Also shown in Figure 1 are external influences that should not be ignored. External influences can include government action, natural occurrences (e.g., storms, floods), and of course competition. Most people directly involved in manufacturing do not normally consider steel mills or refineries or textile mills as ' manufacturing'..

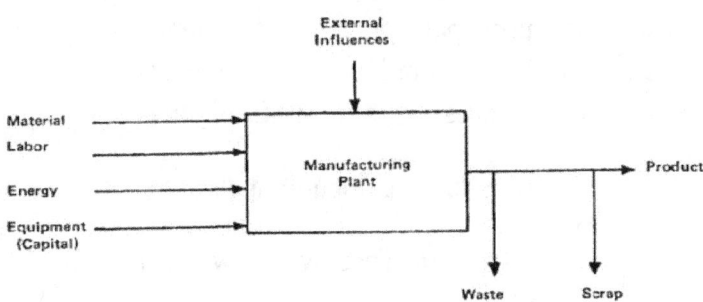

FIGURE 1. Manufacturing system.

3. DEVLOPING AREAS OF A MANUFACTURING SYSTEM

The treatment of the subject matter follows a general rule that technology that became commercial within the past five years is probably still new to many, On the other hand, technologies expected to be available within the next five years need attention. Awareness of the yet-to-come may prepare for quicker technology transfer when this do become current. The focus of the technologies appropriate for manufacturing is on mechanical products and their components. However, a great deal of overlap is expected with all manufacturing industries. A schematic outline of the topic, which requires support and manufacturing development tools, is given in Figure.2. For uniformity in presenting batch manufacturing technology features in this text, the individual operation are divided into one primary and three secondary processes. The principal process is supported by preprocess finishing operations, and the generally grouped secondary processes, as shown in Figure.3.

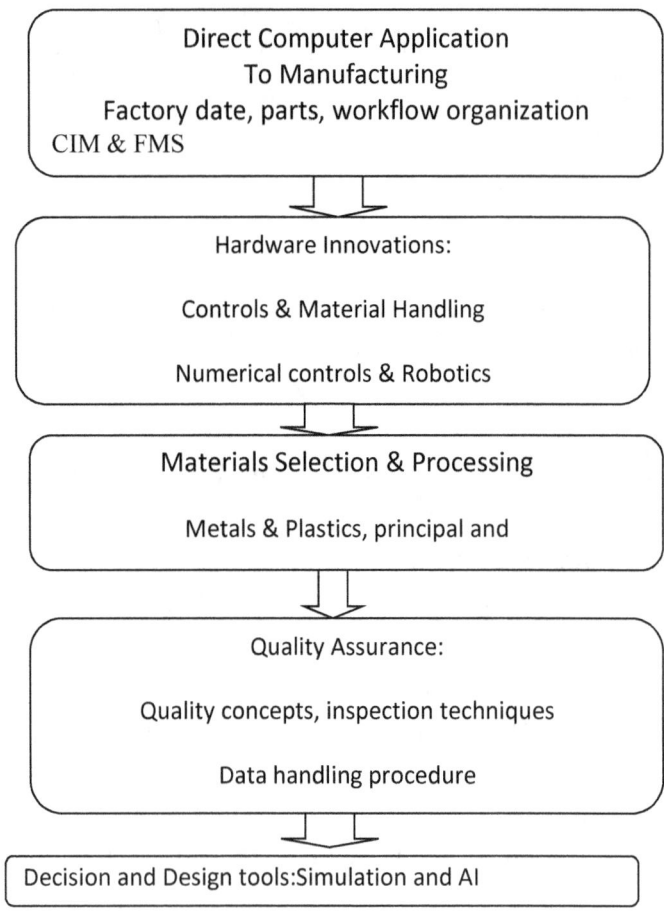

Figure.2. Outline of the various topics, which requires intelligent manufacturing development tools.

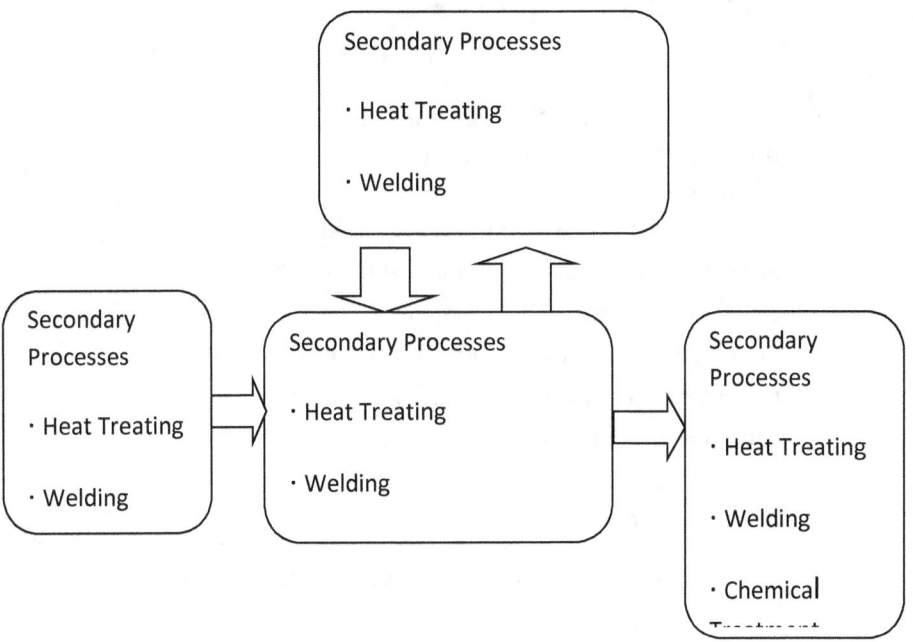

Figure.3 Extended View of machining

4. INTELLIGENT MANUFACTURING SYSTEM

An intelligent Manufacturing process has the ability to self regulate and/or self control to manufacture the product within the design specifications. An integrated concept with factories of future, where products are produced in an artificial life environment adds value to this. Researchers working on implementation of EXPERT

SYSTEM,, have come out with this concept of adding intelligence.

Intelligent manufacturing can be achieved in three basic ways.

❖ Existing manufacturing processes can become intelligent by monitoring and controlling the state of the manufacturing machine.
❖ Existing processes can be made intelligent by adding sensors to monitor and control the state of product being processed.
❖ New processes can be intelligently designed to produce parts of desired quality without the need of sensing and control of the process.

Intelligent Manufacturing system

    A. Uses technology which can minimize the use of human Brain

    B. Regulation for product mix and priority production, self regulated.
    C. Self controlled operations with automatic feedback mechanism.
    D. Monitoring and control of the manufacturing machine.

    E. Monitoring and controlling the state of product being processed.
    F. New processes with intelligence can be made to produce parts of

desired quality without the need of sensing and control of process.

5. TOOLS USED FOR INTELLIGENT MANUFACTURING

Following are the tools generally used in intelligent manufacturing:

　　i.　Fuzzy logic
　ii.　Genetic Algorithms
　iii.　Neural Net Works
　iv.　Case tools
　　v.　Simulation Algorithms.

6. AI TECHNIQUES AND COMPONENTS OF AN INTELLIGENT MANUFACTURING SYSTEMS

This section contains a very brief and much simplified outline of the main AI techniques and the components of a simplified model of an intelligent manufacturing system.

AI Techniques

    A. Knowledge based systems
    B. Neural networks
    C. Fuzzy logic
    D. Genetic algorithms
    E. Case-based reasoning

7. COMPONENTS OF AN INTELLIGENT MANUFACTURING SYSTEM

As mentioned in the previous section, the manufacturing process is a complex one and can be decomposed into several components. Rao et al. (1993) decomposed intelligent manufacturing systems into the following components:

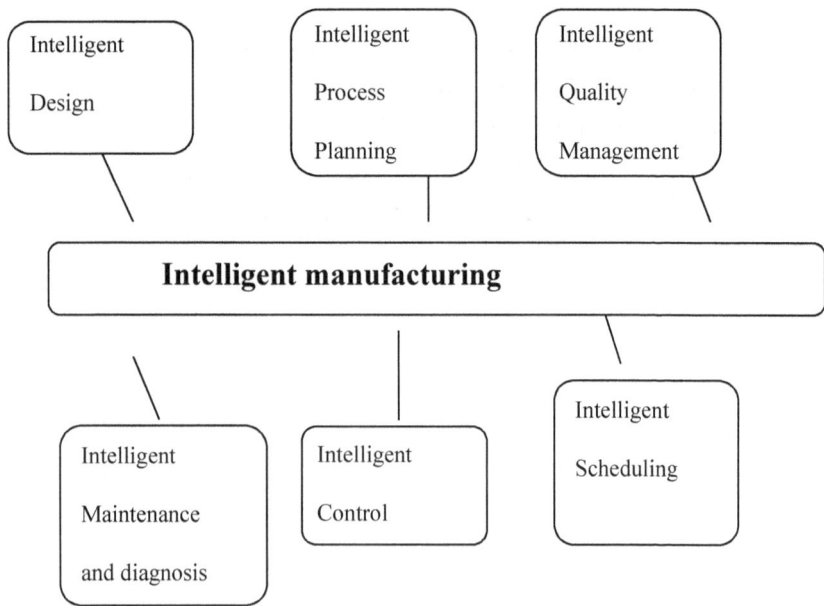

Figure 3: Components of an intelligent manufacturing systems (adapted from (Rao, 1993)

8. USE AND APPLICATION OF AI IN SPECIFIC MANUFACTURING AREAS

The use of AI is increasing very rapidly in manufacturing. Manufacturing will change more in next 15 years than it has in the last 75, largely due to the application of computer-aided technology that has been developed during the last 10 yr. The tools of this technology include:

a) Computer-aided design (CAD)

b) Generative process planning

c) Robotics / material handling

d) Computer-aided manufacturing (CAM)

AI is indicated as the technology that will tie these tools together. The generic application of AI, including the following items directly related to manufacturing:

1) Fault diagnosis and repair (machines and systems)

2) Operation of machines and complex systems

3) Management (Planning, scheduling, and monitoring)

4) Design (systems, equipment, intelligent design aids, and inventing)

11. CONCLUSIONS

In the work, a certain conception of designing intelligent systems for enterprise management was presented. Based on the conception, a methodology of creating the IMS is being developed based on the integration of artificial intelligence technologies with exact methods, well-known in the decision making theory, as well as with simulation modeling methods. The approach proposed will open up a possibility to build an IMS of open structure, combining existing information systems with the information sub-systems in production engineering using artificial intelligence technologies in order to create an integrated environment for comprehensive solving of decision making problems in the system of intelligent manufacturing. Intelligent manufacturing is the most promising and future-oriented of production system developments aiming at further automatization, optimization and integration of manufacturing processes.

REFFERENCES

[1] Deb S.R and Chattopadhayaya, Proc. 3rd SERC school on *Advanced Mfg.*

Tech. Jadavpur Calcutta, 1997

[2] Kumar Surender, *Industrial Robots and CIM*, Oxford and IBH Publication Co,New Delhi.

[3] Kumar Surender and Jha, A.K, *Computer Aided Design and Manufacturing*,Dhanpat Rai and Co. Ltd., New Delhi.

[4] Acosta, L., Marichal, G.N., Moreno, L., Rodrigo, J.J, Hamilton, A., Mendez, J.A (1999), *"A robotic system based on neural network controllers, Artificial Intelligence in Engineering*, 13(4), pp. 393-398

- Several information & pictures are taken from these following websites:
    - www.sail.co.in
    - www.wikipedia.com
    - www.2200blogspot.com
    - www.durgapursteelplant.com
    - www.sail.co.in

 **P.Jana** born in India 1990. Obtained his Bachelor's degree in Production Engineering, from Haldia Institute of Technology, during 2008-2012. & Master's degree from West Bengal University of Technology in the Industrial Engineering & Management during 2012-2014.He is having More than 01 year industrial experience in I.O.C.L and having 05 international journals / Conference papers. He also obtained his professional qualification on NDT ASNT (The American Society for Non-destructive Testing) Level-II (UT, DPT, MPT, RT). He is also a writer of various engineering books.

www.ingramcontent.com/pod-product-compliance
Lightning Source LLC
Chambersburg PA
CBHW071108240526
45469CB00006BD/2387